All the relativistic temporal paradoxes are completely false

CARLO MARIA PACE

ALL THE RELATIVISTIC TEMPORAL PARADOXES ARE COMPLETELY FALSE

Title | All the relativistic temporal paradoxes are completely false
Author | Carlo Maria Pace

ISBN | 978-88-92616-06-6

Youcanprint Self-Publishing
Via Roma, 73 - 73039 Tricase (LE) - Italy
www.youcanprint.it
info@youcanprint.it
Facebook: facebook.com/youcanprint.it
Twitter: twitter.com/youcanprintit

GENERAL INTRODUCTION

In this treatment I demonstrate that all the temporal paradoxes which we meet in the common treatment of relativistic physics are completely false.

In the first chapter, I, by starting from the relativisticly correct definition of time, demonstrate the falsity of the Twin Paradox by means of proving from every relativistic point of view that in no case is one twin not so old as the other twin. In particular, I demonstrate that the Twin Paradox arises erroncously from the use of a relativisticly incorrect definition of time, where this definition is relativisticly incorrect because it is equal to a relativisticly correct definition of time except for the substitution of the phenomena that are motionless relatively to the used spatial reference frame with the analogous phenomena that are in motion relatively to the used spatial reference frame, where these phenomena are the phenomena by means of which time is measured. Moreover, I demonstrate that the so-called "proper" times of objects that are considered in motion are not true times from the relativistic point of view because these so-called "proper" times are not the times of relativisticly correct reference frames[1].

---

[1] In 1994 I have submitted a previous version of the first chapter of this work in the form of an article to the scientific journal *Physical Review A*, the editors of which in a very short time have rejected my article only for the fact that my article was against the traditional position of the academic physicists. In 2016 I have submitted another previous version of the first chapter of this work in the form of an article to the scientific journal *Foundations of Physics*, the editors of which in a very short time have rejected my article without any true explanation. In the same year I have submitted a further previous version of the first chapter of this work in the form of an article to the website *Arxiv.org*, the moderators of which have rejected my article only because I was not a professional member of the scientific community.

In the second chapter, I demonstrate the impossibility of the time travels even in the field of application of the General Theory of Relativity. In particular, I show that the gravitational fields cannot change the measure of time. Moreover, I, by starting from the relativisticly correct definition of time, demonstrate in general the impossibility of the temporal paradoxes, both in the field of application of the Special Theory of Relativity and in the field of application of the General Theory of Relativity.

In the third chapter, I demonstrate the falsity of the temporal paradoxes (that are) based on the two Einsteinian Theories of Relativity also from the fact that we can use two Lorentzian Theories instead of the two Einsteinian Theories of Relativity. In particular, I, by starting from the traditional concept of time, demonstrate that, in general, all the relativistic temporal paradoxes are based only on a substitution, in the two Einsteinian Theories of Relativity, of the traditional time with something else that in a deceptive way is called time as the traditional time.

Finally, in the general conclusion I underline that both the velocities of physical bodies (field of application of the Special Theory of Relativity) and the gravitational fields (field of application of the General Theory of Relativity) cannot change the measure of time.

CHAPTER I
ON THE DEFINITION OF TIME IN THE SPECIAL THEORY OF RELATIVITY: THE FALSITY OF THE TWIN PARADOX

1 Introduction

The principal aim of this chapter is to demonstrate the falsity of the Twin Paradox, by starting from the definition of time according to the Special Theory of Relativity[2]. In other words, this chapter demonstrates, according to the Special Theory of Relativity, that, when the two twins meet again, contrary to the Twin Paradox, they are equally old, and that this is true for every case of relative motion between them during their separation.

---

[2] Cf. A. EINSTEIN, *On the Electrodynamics of Moving Bodies*, Introduction; § 1 in LORENTZ H. A. – EINSTEIN A. – MINKOWSKI H. – WEYL H., *The Principle of Relativity*, Dover Publications, New York 1952, pp. 37-40.

2 The simplest case (that is, the case in which we do not consider the necessary times of acceleration)

Let us designate the twin that remains on Earth as twin 1 and the twin that travels in the rocket as twin 2.

Moreover, for simplicity, we suppose initially that twin 2 moves, with respect to the reference frame that is integral with the Earth and that is considered to be inertial, according to the following relations:

$$v_x = 0 \qquad \text{for} \qquad t \leq 0$$

$$v_x = v \qquad \text{for} \qquad 0 < t < \frac{T}{2}$$

$$v_x = 0 \qquad \text{for} \qquad t = \frac{T}{2}$$

$$v_x = -v \qquad \text{for} \qquad \frac{T}{2} < t < T$$

$$v_x = 0 \qquad \text{for} \qquad t \geq T$$

(For simplicity, in these relations I have neglected the necessary times of acceleration).

Therefore twin 1 sees, relatively to his inertial reference frame (ct, x) that is integral with himself, this Minkowskian graph (Fig. 1):

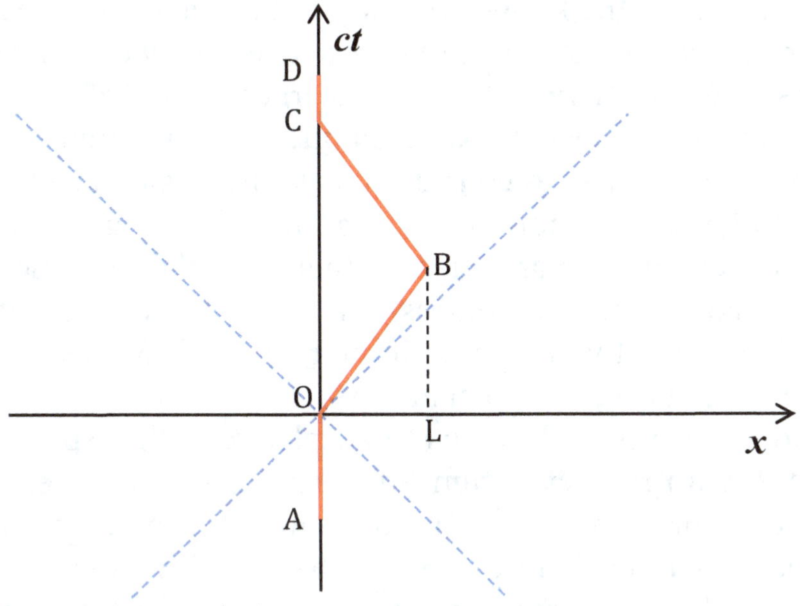

Fig. 1: In this graph twin 1 moves, as time elapses, along the axis *ct* from A to D, while twin 2, as time elapses, follows the crooked line in red: AOBCD.

In this reference frame of twin 1, the time that is elapsed between O and C is equal to the time of the journey there and back of the rocket of twin 2, and this time is equal to $T = \dfrac{2L}{v}$. In particular, the elapsed time is equal to $\dfrac{L}{v}$ both for the outward journey and for the return journey.

The definition of the interval of time in the Special Theory of Relativity is exactly equivalent to what I have done: I have used the rocket at the known velocity v, instead of a ray of light in the vacuum at the known velocity c, between two fixed points of the reference frame at a known distance L between them[3].

As far as twin 2 is concerned, for effecting any measurement, he must choose a unique reference frame for each measurement and has to consider it (that is, this unique reference frame) to be always motionless from the beginning to the end of the measurement; in particular, in this case, the reference frame has to be considered motionless at least from the moment of the separation of the two twins until their reunion for measuring the elapsed time during their separation.

Now, once twin 2 begins to accelerate with respect to twin 1, twin 2, with respect to himself, sees other accelerated bodies, of which, for explaining the dynamics, twin 2 must either introduce "fictitious" forces, such that the Principle of Inertia remain valid in the reference frame that is integral with himself, or admit that he, with respect to himself, sees some "apparent" accelerations that are really due to the fact that he is moving with accelerated motion. In this second case twin 2, in saying that he is accelerating, is referring to the reference frame that he had before beginning to accelerate: the reference frame of twin 1. In other words, in this second case, 1) twin 2, during the measurement, considers the Earth to be motionless as does twin 1; 2) twin 2 sees the world with the same physical laws of twin 1 (since twin 2, in order not to modify the physical laws, says that he is moving with accelerated motion); 3) twin 2 measures times with the same definition of time as before the departure, and therefore, formally, with a definition of time that is equal to that of twin 1 (for example, by means of a ray of light that goes forward and back in the vacuum between two fixed points of the reference frame at a known distance between them). Now, because of the equality of the physical laws (that are) observed

---

[3] Cf. A. EINSTEIN, *On the Electrodynamics of Moving Bodies*, Introduction; § 1 in LORENTZ H. A. – EINSTEIN A. – MINKOWSKI H. – WEYL H., *The Principle of Relativity*, Dover Publications, New York 1952, pp. 37-40.

by the two twins, (in particular the same velocity of light in the vacuum c with respect to the Earth), time elapses equally for both of them, who, likewise, have the same temporal origin $t = 0$ at the moment of their separation. Therefore the two reference frames of the two twins in this case coincide; for this reason the elapsed times are the same for both twins and are equal to the time that we have seen elapse for twin 1.

The authors who say that the Twin Paradox is true use as reference frame of twin 2 a manifestly incorrect reference frame, because they say that twin 2 cannot consider himself to be motionless during the journey[4], and that, therefore, twin 2, as the spatial reference frame with respect to which he sees the spatial displacements, uses the spatial reference frame in which he was before the journey: the spatial reference frame of the Earth (that is, that of twin 1); but, as far as the measure of time is concerned, these authors make twin 2 use phenomena that are integral with himself and therefore in motion at a velocity $v(t)$ with respect to the spatial reference frame that is used by twin 2, that is to say with respect to the spatial reference frame that is integral with the Earth, instead of the analogous motionless (with respect to the Earth) phenomena in a definition of the measure of time that is a relativisticly correct definition only by means of the use of the analogous motionless (with respect to the Earth) phenomena. Therefore, these measurements of time are substantially in contrast with those (that are) obtained using the relativistic definition of time in an inertial reference frame, where this definition is given by the time that a ray of light takes for going forward and back in the vacuum, at the known velocity c, over a known distance between two fixed points in the used inertial reference frame[5]. In other words, these authors do not consider

---

[4] Cf. RINDLER p. 46; ANDERSON p. 176; OHANIAN – RUFFINI p. 176.

[5] Cf. A. EINSTEIN, *On the Electrodynamics of Moving Bodies*, Introduction; § 1 in LORENTZ H. A. – EINSTEIN A. – MINKOWSKI H. – WEYL H., *The Principle of Relativity*, Dover Publications, New York 1952, pp. 37-40.

the fact that they see the phenomena in motion happen more slowly than the analogous motionless phenomena by a factor $\gamma \equiv \dfrac{1}{\sqrt{1-\dfrac{v^2}{c^2}}}$, and that, therefore, they must multiply the times

that are measured using moving phenomena by γ for obtaining the relativisticly correct times, that they would measure by means of using the analogous motionless phenomena. Therefore, in the relativisticly correct definition of time, they substitute the motionless phenomena with the analogous moving phenomena, and so obtain a relativisticly incorrect definition of time, by a

factor γ, which gives rise to the Twin Paradox, that is therefore evidently false.

Whereas twin 1 uses, as spatial reference frame, a reference frame that is integral with the Earth, and, as measure of time, uses phenomena that take place on the Earth (that is, phenomena that are motionless with respect to the Earth) equivalently to the definition of time by means of measurements that are effected by means of a ray of light that goes forward and back in the vacuum, at the known velocity c, between two fixed points in the reference frame at a known distance between them, which definition has been given by A. Einstein in the Special Theory of Relativity[6].

Therefore, the so-called "asymmetry" between the two reference frames that are used by the two twins is due only to the fact that twin 1 uses a true inertial reference frame, in conformity with the Special Theory of Relativity, while twin 2 uses a reference frame in which time is incorrectly defined, or rather in which time is defined in a way contrary to the definition of time, which (definition) has been given by A. Einstein in the Special

---

[6] Cf. A. EINSTEIN, *On the Electrodynamics of Moving Bodies*, Introduction; § 1 in LORENTZ H. A. – EINSTEIN A. – MINKOWSKI H. – WEYL H., *The Principle of Relativity*, Dover Publications, New York 1952, pp. 37-40.

Theory of Relativity[7], since in this reference frame time is measured by means of moving phenomena and not by means of motionless phenomena, while the relativisticly correct definition of time is by means of motionless phenomena and not by means of moving phenomena.

In particular, we have, precisely from this incorrect definition, the factor $\gamma(t) \equiv \dfrac{1}{\sqrt{1-\dfrac{v(t)^2}{c^2}}}$, which erroneously gives rise to the Twin Paradox. To see this in a simple manner, let us consider the case in which one uses this incorrect definition of the reference frame of twin 2 in a case where twin 1 is on the Earth and twin 2 in a rocket at an always constant velocity v with respect to the Earth, during the measurements. Thus, twin 2, since he considers the Earth to be motionless, himself to be in motion, and time to be measured by means of moving phenomena that are integral with himself at the velocity v with respect to the Earth which is integral with the spatial reference frame of twin 2, measures a time shorter by a factor $\gamma \equiv \dfrac{1}{\sqrt{1-\dfrac{v^2}{c^2}}}$ than the time that he "sees" measured from the Earth. Therefore, the measures of the two used reference frames are in agreement in saying that twin 2 has led a shorter life. But this is manifestly incorrect, because the two true reference frames, that are relativisticly correct, are perfectly symmetric between them since the true reference frame that is integral with twin 2 is that in which twin 2 is motionless and the Earth is in motion at the velocity $-v$. Or rather, exchanging the roles of the two twins, we could turn over the above reasoning completely, thus arriving at

---

[7] Cf. A. EINSTEIN, *On the Electrodynamics of Moving Bodies*, Introduction; § 1 in LORENTZ H. A. – EINSTEIN A. – MINKOWSKI H. – WEYL H., *The Principle of Relativity*, Dover Publications, New York 1952, pp. 37-40.

an absurdity since in this case twin 2 would be the older, while in the previous case twin 2 would be the younger.

In other words, we can say that this absurdity arises from the fact that the upholders of the Twin Paradox choose as a reference frame of twin 2 a four-dimensional reference frame in which the time axis (that is, the temporal axis) is not perpendicular to the spatial axes, since in this reference frame the projection of the time axis on the spatial axes is not nothing, as one sees from the fact that twin 2, by means of moving along the time axis of the reference frame that he uses, also moves in space relatively to the same reference frame. Consequently, in this reference frame of twin 2 the velocity of light in the vacuum is not isotropic, because the time axis that is used is not symmetric with respect to the light cone! Obviously, this reference frame is not a correct reference frame from the relativistic point of view and the Twin Paradox arises just from this error, therefore we have to consider the Twin Paradox incorrect.

Instead, in the case in which we introduce "fictitious" forces (as mentioned above), twin 2 performs measures with respect to an inertial reference frame (ct', x') that is motionless from the beginning to the end of the measurement and that is integral with himself (that is, with twin 2), for which twin 2 sees this Minkowskian graph (Fig. 2):

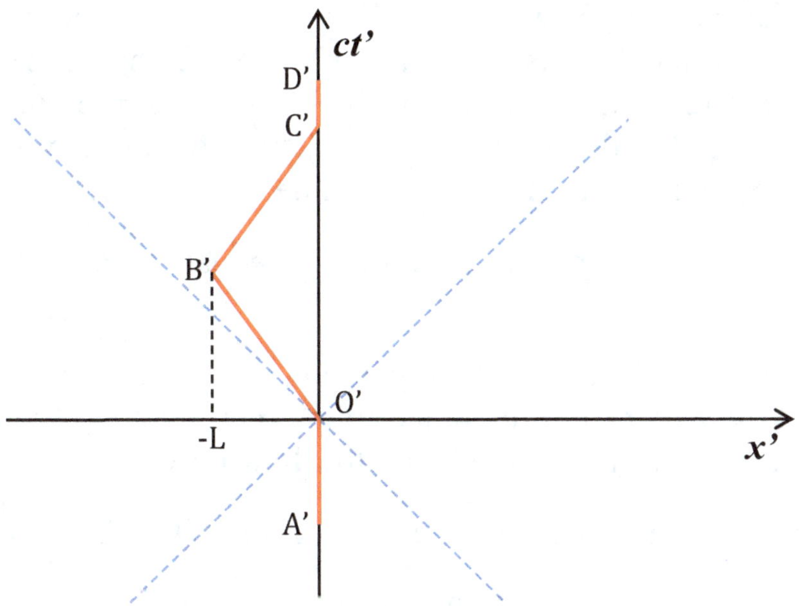

Fig. 2: Twin 2 moves, as time elapses, along the axis **ct'** from A' to D', while twin 1 moves, as time elapses, along the crooked line in red: A'O'B'C'D'.

The time (that is) elapsed from O' to C' is, also here, equal to $T = \dfrac{2L}{v}$. In particular the elapsed time is equal to $\dfrac{L}{v}$ both for the outward journey and for the return journey. Therefore, when the two twins meet again, the two twins are equally old, because each of them, during the separation, has led a life of a length that is equal to the same time T.

The case is different for the so-called "proper" time of a twin that is considered in motion, because then the proper time is equivalent to the elapsed time that one observes (from his own reference frame) in a different reference frame that is in motion with respect to his own. Twin 1, looking from the reference frame that is integral with himself, sees[8] as the time axis of the reference frame that is integral with twin 2 the crooked line AOBCD, and, as the measure of the time in the reference frame that is integral with twin 2, sees the length of this crooked line. Now the length of this crooked line, during the separation between the two twins, has a contraction equal to $\gamma \equiv \dfrac{1}{\sqrt{1-\dfrac{v^2}{c^2}}}$ relatively to the length of the projection of the same crooked line on the time axis of the reference frame that is integral with twin 1, because of which twin 1, for the line OBC, sees (from the reference frame that is integral with himself) a time $\tilde{T} = \dfrac{T}{\gamma}$ pass in the reference frame that is integral with twin 2. Analogously twin 2, during the separation between the two twins, sees (from the reference frame that is integral with himself) a time $\dfrac{T}{\gamma}$ pass in the reference frame that is integral with twin 1[9].

Therefore, these contracted measures are useful as a measure of the "distortion" with which the relative velocity makes us see phenomena in motion, but they have nothing to do with the correct definitions of time. In particular, each of the two twins has seen in a "distorted" way phenomena which were integral with the other reference frame, in consequence of the relative velocity, but this does not contrast with the fact that, when the two twins meet again, the two twins are equally old.

An analogous case occurs between two inertial reference frames, where one of these inertial reference frames is in motion at a velocity v with respect to the other, since from each of them we see phenomena, which are integral with the other reference

---

[8] Cf. Fig. 1.
[9] Cf. Fig. 2.

frame, happen more slowly by a factor $\gamma \equiv \dfrac{1}{\sqrt{1-\dfrac{v^2}{c^2}}}$; but each

of the two reference frames, in reality, measures time by means of phenomena that are integral with itself, because of which the phenomena, that are integral with one of the two reference frames, in motion at a velocity v with respect to the other reference frame, are not used by this other reference frame for measuring time, and therefore the times (that are) contracted by a

factor γ are not true times from the relativistic point of view, because they are not times that are measured as times must be measured in the Special Theory of Relativity, both in the first and in the second reference frame. On the other hand, we could use phenomena in motion for measuring time, provided that we allow for the fact that we see these phenomena happen more slowly by

a factor γ, thus returning to a measure (that is) equivalent to the

relativisticly correct measure, without an erroneous factor γ. By eliminating the measures that are not relativisticly correct and

that have the erroneous factor γ, the relativisticly correct measures are perfectly equal, since following the reasoning, that we have developed for the two twins, relatively to the outward journey only we see that the times are equal in the two four-dimensional inertial reference frames, which are correct from the relativistic point of view and which are integral respectively with the two twins. On the other hand, these reference frames are both inertial relatively to the outward journey because we have neglected the times of acceleration, even if these times, in reality, are necessary. The only variation is that we use, for the definition of time in a reference frame, the time that is taken by an object at a known velocity v for going between two fixed points of the reference frame at a known distance L between them; but this definition is also perfectly analogous and equivalent to the relativistic definition of time, because in the Special Theory of Relativity A. Einstein, by means of a similar method, synchronizes the clocks of an inertial reference frame that are placed in all the points of space[10].

3 The general case (that is, the case in which we consider realistic accelerations)

Moreover, we note that, also in the case of the two twins with realistic accelerations, by means of dividing the elapsed time into infinitesimal short intervals of time, we, at the limit, in every infinitesimal short interval, can consider that the two twins are integral respectively with two inertial reference frames, where one of these inertial reference frames is at a constant velocity v with respect to the other.

Now, for every infinitesimal short interval, that is to say for every pair of these pairs of inertial reference frames, the times that elapse in the two inertial reference frames of the pair, as we have seen, are equal and therefore, integrating all the short intervals, we see that the total elapsed time is the same both for twin 1 and twin 2. This result is also evident because of the fact that every pair of inertial reference frames is perfectly symmetric in the two inertial reference frames and therefore, even if we integrate, the two twins find themselves to use two groups of inertial reference frames, where these groups are perfectly symmetric between them from the relativistic point of view. Thus, to conclude that the total time (that is) elapsed for one twin is different from the total time (that is) elapsed for the other twin would be absurd from the relativistic point of view.

---

[10] Cf. A. EINSTEIN, *On the Electrodynamics of Moving Bodies*, Introduction; § 1 in LORENTZ H. A. – EINSTEIN A. – MINKOWSKI H. – WEYL H., *The Principle of Relativity*, Dover Publications, New York 1952, pp. 37-40.

On the other hand, the reasoning that we have applied to a particular motion[11] can be generalized to any motion whatever.

Therefore, if twin 1 sees things from his own inertial reference frame in this manner (Fig. 3):

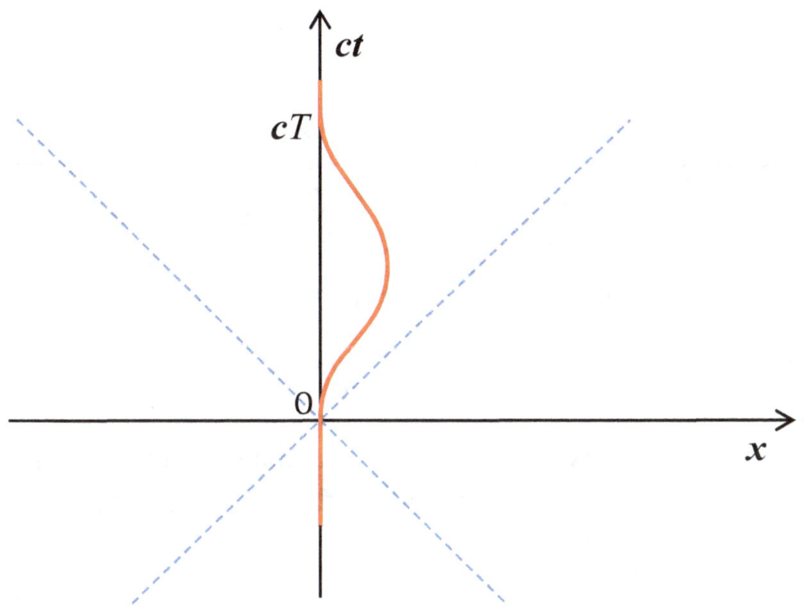

Fig. 3: Twin 1 moves, as time elapses, along the time axis, while twin 2 moves, as time elapses, along the trajectory in red.

Then twin 2 (if twin 2 uses the reference frame of twin 1, the result is banal) from his own reference frame that is integral with the rocket (and that he considers motionless and inertial) sees (Fig. 4):

---

[11] Cf. Fig. 1 and Fig. 2.

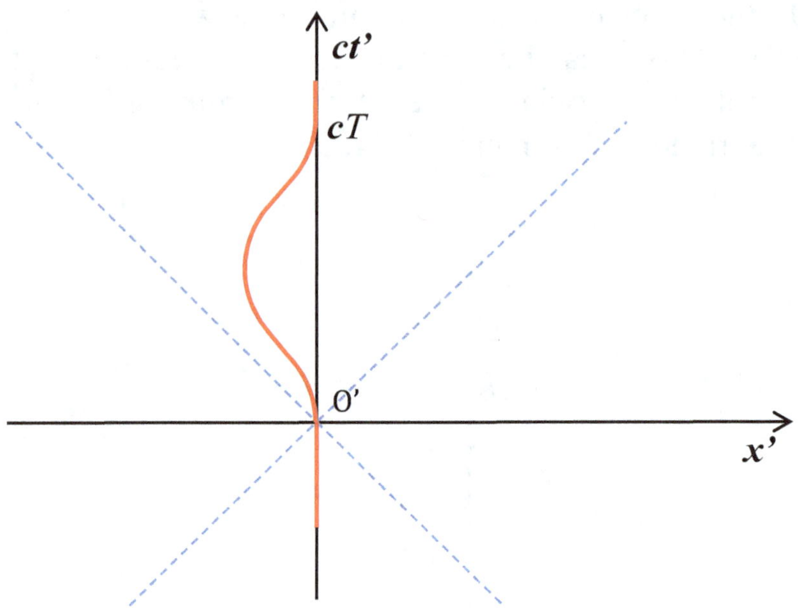

Fig. 4: Twin 2 moves, as time elapses, along the time axis, while twin 1 moves, as time elapses, along the trajectory in red.

The only difference between them is the change of the sign of the relative velocity (that is, the direction of the journey). Now, since the measurements of time are not due to the orientation of the velocity, that is to say to the direction of the journey, we see that the two reference frames are perfectly equivalent and symmetric from the point of view of the temporal measurements. Therefore, every twin measures the time of the separation between the two twins equal to T.

Moreover, each twin, looking from his own reference frame, sees the time of the other reference frame elapse more slowly. In particular,

$$\tilde{T} = \int_0^T \sqrt{1 - \frac{v^2(t)}{c^2}}\, dt$$

is the time that each twin, looking from his own reference frame, sees elapse in the other reference frame during the separation.

Since $\tilde{T} < T$ we have that each twin has seen in a "distorted" way the actions of the other twin; in particular, each twin has seen the actions of the other twin slowed by a factor $\gamma(t)$. But the final result does not change; in fact during the separation each twin has measured the same time, and therefore the two twins are equally old when they meet again, since they have led an equally long life, that is to say a life of the same length of time. Also in this case, we can, obviously, divide all into infinitesimal short intervals of time, and can, at the limit, in every short interval, consider that the two twins are integral respectively with two inertial reference frames, where one of these inertial reference frames has a constant velocity v with respect to the other. Repeating the already explained reasoning, we see that the times measured in the two groups of reference frames, where these groups are those that are used respectively by the two twins, are equal.

Clearly, to combine the spatial reference frame that is integral with the Earth, with the time axis that is given by the trajectory of twin 2, who is not motionless spatially with respect to the Earth, is, also here, absurd from the relativistic point of view. Therefore, the reasoning that starts exactly from this erroneous combination, and that is unfortunately used by a lot of books in their treatment of the Twin Paradox, is obviously incorrect.

4 The case of the decay of elementary particles in motion

Likewise, we see phenomena, for example the decay of elementary particles, in motion at a velocity v, happen γ (where $\gamma \equiv \dfrac{1}{\sqrt{1-\dfrac{v^2}{c^2}}}$) times more slowly than the analogous motionless phenomena, but these factors γ are always relative to the ratio between the times that are measured in a relativisticly correct reference frame and the times that are measured in a relativisticly incorrect reference frame; or rather relative to the ratio between times that are measured in the same inertial reference frame, once correctly and another time incorrectly from the point of view of the relativistic definition (that has been) given by A. Einstein for the measure of time[12].

5 The case of the atomic clocks that have been placed on aeroplanes

In the experiment (that has been performed in 1971 by J. C. Hafele and R. E. Keating) of the atomic clocks that have been placed on aeroplanes[13], the used spatiotemporal reference frame, practically, was always the reference frame that is integral with the Earth, and therefore (in this experiment) the measures (that have been) taken from clocks (that were) motionless relatively to the used reference frame have been compared with the measures

---

[12] Cf. A. EINSTEIN, *On the Electrodynamics of Moving Bodies*, Introduction; § 1 in LORENTZ H. A. – EINSTEIN A. – MINKOWSKI H. – WEYL H., *The Principle of Relativity*, Dover Publications, New York 1952, pp. 37-40.
[13] Cf. CRAIG pp. 67-68; EISBERG – RESNICK p. A-9.

(that have been) taken from clocks in motion relatively to the used reference frame: and in these latter clocks, obviously, the phenomena have been seen to happen more slowly. Therefore, practically, the authors of this experiment have compared measures (that have been) effected by means of well calibrated instruments (since they were motionless relatively to the used reference frame) with measures (that have been) effected by means of not well calibrated instruments (since they were in motion relatively to the used reference frame). In other words, in this case the authors of this experiment have compared measures (that have been) effected in a relativisticly correct reference frame (when the clocks were considered motionless) with measures (that have been) effected in a relativisticly incorrect reference frame (when the clocks were considered in motion).

6 About the treatment (that has been) made by J. R. Pierce

With regard to the treatment (that has been) made by J. R. Pierce of the Twin Paradox[14], we see that the times (that are) measured relatively to the reference frame of twin 2, that is to say relatively to the reference frame that is integral with the rocket, are incorrect. In fact, in the first place, Pierce has measured time by means of a fixed point (that is the rocket) between two points in motion, contrary to the definition (that has been) given by A. Einstein, which is by means of two fixed points and one in motion[15]: in other words, Pierce too has measured time by means of moving phenomena instead of by means of motionless phenomena. Moreover, Pierce has made the frequency, that is

---

[14] Cf. PIERCE pp. 1053-1061.
[15] Cf. A. EINSTEIN, *On the Electrodynamics of Moving Bodies*, Introduction; § 1 in LORENTZ H. A. – EINSTEIN A. – MINKOWSKI H. – WEYL H., *The Principle of Relativity*, Dover Publications, New York 1952, pp. 37-40.

received by the rocket, of a radiation that has been previously emitted from the Earth, change instantaneously because of the Doppler effect. Now, this change is conceivable only by imagining that the velocity of the rocket has changed, but to imagine this, in the reference frame that is integral with the same rocket, (as Pierce has done) is absurd. In other words, Pierce too has made the error of using for twin 2 a reference frame in which the time axis is not perpendicular to the spatial axes.

7 A general observation on the measure of times

In conclusion, we make a final observation. In general, in physics, two measures can be compared only if they (the two measures) are relative to a same reference frame. Now, for every reference frame that is consistent with the Special Theory of Relativity, we see that the two twins, in the moment in which they separate and in the moment in which they meet again, are respectively in two points of space-time $(t, \vec{x})$ and $(t', \vec{x'})$, and that therefore the time that is elapsed between these two events (that is, the events of the beginning and of the end of the separation between the two twins) is equal, by definition, to $\Delta t = t' - t$ and is the same for both the two twins, since in every reference frame that is consistent with the Special Theory of Relativity there is one and only one time axis, and this axis is valid for all bodies that are either motionless or in any motion whatever, and, by definition, the time that is elapsed between two events is given by the difference of the values of the temporal coordinates of the two events. Therefore, in every reference frame that is consistent with the Special Theory of Relativity we indisputably see that the correctly defined times are equal for the two twins, and that, therefore, from every relativistic point of view, they age at the same rate.

Unfortunately many people erroneously use different time axes respectively for bodies that are in motion in different ways (as functions of time) while the spatial reference frame is always the same, that is to say they use, in a same reference frame, clocks that are in motion in different ways (as functions of time) respectively for bodies that are in motion in different ways (as functions of time)[16]. They, in conformity with this incorrect use, no longer have a fixed temporal distance between two spatiotemporal points that are fixed relatively to the reference frame. They have, rather, an infinite number of temporal distances between these two points, in conformity with the infinite number of ways with which one can go from the first to the second point, since they use in fact an infinite number of time axes, where every time axis is used for those objects that are in motion in a certain way (as a function of time). Now, this is completely absurd, since, in every relativisticly correct reference frame, there is always, by definition, one and only one time axis (that is perpendicular to the used spatial axes).

8 Conclusion

Finally, the Twin Paradox arises exactly from this error: that is, from considering one of the erroneous reference frames as relativisticly correct, not allowing the reverse of the result by means of inverting the roles of the two twins. This error becomes even more evident when we consider that the two twins in every moment are integral respectively with two inertial reference frames, where one of these two inertial reference frames has a

---

[16] Cf. MISNER – THORNE – WHEELER pp. 164.167.1054-1055; DAVIES pp. 55-65.74-75; GOTT pp. 69-72; CRAIG pp. 47-59; T. VAN FLANDERN in CRAIG – SMITH pp. 221-226.

constant velocity v with respect to the other, because, by means of applying the same erroneous considerations that give rise to the Twin Paradox, we find ourselves in an absurdity once we reverse the roles of the two inertial reference frames, that are symmetric between them from every relativistic point of view.

CHAPTER II
ON THE DEFINITION OF TIME
IN THE GENERAL THEORY OF
RELATIVITY:
THE IMPOSSIBILITY OF
THE TIME TRAVELS

1 Introduction

The principal aim of this chapter is to demonstrate the impossibility of the time travels in the field of application of the General Theory of Relativity, by starting from the definition of time according to the Special Theory of Relativity[17]. In fact it is not possible to use a concept of time in the General Theory of Relativity (which concept is) without any relation with the definition of time in the Special Theory of Relativity, otherwise we would call time a thing that is not time in a real sense.

I have already demonstrated (in the first chapter), in the field of application of the Special Theory of Relativity, the falsity of the famous Twin Paradox (and therefore also of the analogous temporal paradoxes in the field of application of the Special Theory of Relativity), by starting from the definition of time according to the Special Theory of Relativity[18]. Now, in this

---

[17] Cf. A. EINSTEIN, *On the Electrodynamics of Moving Bodies*, Introduction; § 1 in LORENTZ H. A. – EINSTEIN A. – MINKOWSKI H. – WEYL H., *The Principle of Relativity*, Dover Publications, New York 1952, pp. 37-40.

[18] Cf. A. EINSTEIN, *On the Electrodynamics of Moving Bodies*, Introduction; § 1 in LORENTZ H. A. – EINSTEIN A. – MINKOWSKI H. – WEYL H., *The Principle of Relativity*, Dover Publications, New York 1952, pp. 37-40.

chapter I will demonstrate the falsity of the temporal paradoxes that are peculiar to the General Theory of Relativity, in particular of the paradoxes that are relative to the possibility of time travels in the past.

2 The effects of gravity on clocks

According to the General Theory of Relativity clocks go more slowly when they are in a gravitational field[19]. But the fact that clocks in presence of a gravitational field go more slowly and at the limit stop does not mean that time itself goes more slowly or stops: in fact if we slow down or stop the hands of a clock by means of a mechanic operation made by us, we do not say that time slows down or stops but that the clock because of our intervention does not sign any longer the correct time. In an analogous way clocks sign the correct time in presence of gravitational fields only if we allow for the fact that gravitational fields slow down or stop the clocks, thus returning to a measure (which is) equivalent to the correct measure, that is to say to the measure in the case without gravitational fields.

On the other hand, only in this way we use a definition of time (which definition is) analogous to that of the Special Theory of Relativity, in which theory we do not consider any gravitational field; and we have seen that we must use a definition of time (which definition is) analogous to that of the Special Theory of Relativity also in the field of application of the General Theory of Relativity, if we want speak of time in a real sense.

------

[19] Cf. A. EINSTEIN, *The Foundation of the General Theory of Relativity*, § 22 in LORENTZ H. A. – EINSTEIN A. – MINKOWSKI H. – WEYL H., *The Principle of Relativity*, Dover Publications, New York 1952, pp. 161-162; OHANIAN – RUFFINI pp. 130-137; WEINBERG pp. 79-85; ANDERSON pp. 414-418; HAWKING p. 48; EVERETT – ROMAN pp. 138-139; T. VAN FLANDERN in CRAIG – SMITH p. 213.

Therefore, in conclusion, we can say that time itself is not influenced by gravitational fields, even if clocks go more slowly when they are in a gravitational field, since these clocks are not well calibrated. On the other hand, the well calibrated clocks, by definition, are not influenced by gravitational fields.

3 The influence of gravitational fields on our rate of aging

Instead, with regard to the question if gravitational fields slow down the rate of aging, we must say that in general the rate of aging does not depend on the gravitational forces, even if gravitational forces (that are) too diverse from those that are present on the Earth can damage our health.

In particular, a too big difference of gravitational forces between different parts of our bodies can influence our health in the sense that we can be damaged by a too big difference of gravitational fields between different parts of our bodies: but in this case our life is reduced by gravitational fields and is not increased by them, contrary to the usual exposition of the temporal paradoxes in the field of application of the General Theory of Relativity.

Moreover, too big gravitational forces can damage our health when we are not in free fall, because our body is not able to sustain too big forces on itself without any damage. Of course, also in this case our life is reduced by too big gravitational fields and is not increased by them, contrary to the usual exposition of the temporal paradoxes in the field of application of the General Theory of Relativity.

However, the rate of aging is different from the health, therefore even in the cases in which the health is damaged by gravitational fields the rate of aging can be and is the same.

So, not only in the field of application of the Special Theory of Relativity (that is, in presence of velocities which are close to the velocity of light in the vacuum) but also in the field of application of the General Theory of Relativity (that is, in presence of gravitational fields) when we have the same times in according to well calibrated clocks, in general we also have the same rate of aging for all human beings.

4 The time travels in the past are impossible

4.1 The time travels that are based on the incurvature (due to gravitational fields) of space-time

Once we delete the effects of the gravitational fields on the measure of time for measuring time in a correct way, we see that the time travels in the past that are based on gravitational effects are completely impossible, also in the field of application of the General Theory of Relativity. In fact the alleged time travels in the past, that are based on the incurvature (of space-time) due to gravitational fields, are impossible for the fact that, as we have seen, time must be measured only after that we have eliminated any effect due to gravitational fields. In other words, the presence of gravitational fields cannot change the correct measure of time from a value to another value, and therefore, in particular, cannot change the correct measure of time from a positive value to a negative value, since the correct measure of time does not depend on the presence or the absence of gravitational fields. Therefore, in particular, the time travels, that are based only on wormholes (due to gravitational fields) or black holes, are impossible.

4.2 The time travels that are based on velocities greater than that of light in the vacuum

On the other hand, also the time travels, that are based only on velocities greater than that of light in the vacuum, are impossible, since the presence of persons (or of things or simply of true information) that can have velocities greater than that of light in the vacuum would prove that the Special Theory of Relativity is false and that instead is true a Lorentzian Theory. In fact we could use these superluminal signals (or persons or things) for establishing an absolute reference frame according to a Lorentzian Theory. In this case time would be absolute (in the field of application of the Special Theory of Relativity and hence in the field of application of the General Theory of Relativity) and therefore it would be intrinsically impossible any time travel in the past: in fact, all things would go ahead in this absolute time in the same way and therefore all the temporal paradoxes would be completely impossible!

Moreover, the presence of wormholes would enable velocities greater than that of light in the vacuum and therefore would imply the truth of two Lorentzian Theories instead of the two Einsteinian Theories of Relativity: therefore also for this motive the time travels due only to the presence of wormholes are impossible.

More in general, a time travel in the past would imply the capacity of traveling faster than light in the vacuum, and therefore (would imply) also the falsity both of the Special Theory of Relativity and of the General Theory of Relativity, and the truth of two corresponding Lorentzian Theories, which truth would imply the impossibility of any time travel, since in these two theories there would be an absolute time. In conclusion, the possibility of a time travel in the past would imply the impossibility of any time travel, therefore any time travel in the past is impossible, since its possibility is self-contradictory.

5 A general observation on the measure of times

Moreover, we can make a final observation, which is analogous to that of the first chapter. In general, in physics, two measures can be compared only if they (the two measures) are relative to a same reference frame. Now, for every reference frame that is consistent with the Special Theory of Relativity (and analogously with the General Theory of Relativity), we see that any couple of events is relative to two points of space-time $(t, \vec{x})$ and $(t', \vec{x'})$, and that therefore the time that is elapsed between these two events is equal, by definition, to $\Delta t = t' - t$, since in every reference frame that is consistent with the Special Theory of Relativity (and analogously with the General Theory of Relativity) there is one and only one time axis, and this axis is valid for all bodies that in any gravitational field are either motionless or in any motion whatever, and, by definition, the time that is elapsed between two events is given by the difference of the values of the temporal coordinates of the two events.

Unfortunately many people erroneously use different time axes respectively for bodies that have different velocities (as functions of time) and/or different gravitational fields while the spatial reference frame is always the same, that is to say they use, in a same reference frame, clocks at different velocities (as functions of time) and/or in different gravitational fields respectively for bodies at different velocities (as functions of time) and/or in different gravitational fields[20]. They, in conformity with this incorrect use, no longer have a fixed temporal distance between two spatiotemporal points that are fixed relatively to the reference frame. They have, rather, an

---

[20] Cf. MISNER – THORNE – WHEELER pp. 164.167.1054-1055; DAVIES pp. 55-65.74-75; GOTT pp. 69-72; CRAIG pp. 47-59; T. VAN FLANDERN in CRAIG – SMITH pp. 221-226.

infinite number of temporal distances between these two points, in conformity with the infinite number of ways with which one can go from the first to the second point, since they use in fact an infinite number of time axes, where every axis is used for those objects which have a certain velocity (as a function of time) and are in a certain gravitational field. Now, this is completely absurd, since, in every relativisticly correct reference frame (both in the Special Theory of Relativity and in the General Theory of Relativity), there is always, by definition, one and only one time axis (that is perpendicular to the used spatial axes) and the time of this axis is measured apart from any velocity of objects and/or any gravitational field.

6 The General Theory of Relativity as extension of the Special Theory of Relativity

In conclusion, in the Special Theory of Relativity the temporal paradoxes (and in particular the Twin Paradox) arise exactly from this error: that is, from considering one of the erroneous reference frames as relativisticly correct, not allowing the reverse of the results by means of inverting the roles of the two observers. This error becomes even more evident when we consider that the two observers in every moment are integral respectively with two inertial reference frames, where one of these two inertial reference frames has a constant velocity v with respect to the other, because by applying the same erroneous considerations that give rise to the temporal paradoxes (and in particular to the Twin Paradox), we find ourselves in an absurdity once we reverse the roles of the two inertial reference frames, that are symmetric between them from every relativistic point of view, as we have seen in the first chapter.

Now, the General Theory of Relativity is an extension of the Special Theory of Relativity to the case with the presence of gravitational fields. But, on the other hand, as we have seen, the presence of gravitational fields cannot change the correct measure of time, therefore the temporal paradoxes are erroneous not only in the field of application of the Special Theory of Relativity but also in the field of application of the General Theory of Relativity.

CHAPTER III
THE FALSITY OF THE RELATIVISTIC TEMPORAL PARADOXES IS ALSO PROVED BY THE POSSIBILITY OF USING TWO LORENTZIAN THEORIES INSTEAD OF THE TWO EINSTEINIAN THEORIES OF RELATIVITY

We can see the falsity of the temporal paradoxes that are based on the two Einsteinian Theories of Relativity also from the fact that we can use two Lorentzian Theories instead of the two Einsteinian Theories of Relativity[21].

In fact if we use the corresponding Lorentzian Theories instead of the analogous Einsteinian Theories of Relativity, we do not meet any temporal paradox, since in this case there is an absolute reference frame and therefore an absolute time that is relative to this absolute reference frame: all things go ahead in this absolute time in the same way and therefore all the temporal paradoxes are completely impossible! This clearly shows us that the temporal paradoxes are not based on something real but only on a substitution, in the two Einsteinian Theories of Relativity, of

---

[21] Cf. CRAIG pp. 105-242; W. L. CRAIG in CRAIG – SMITH pp. 11-49; Q. SMITH in CRAIG – SMITH pp. 75-76.82-83.90; T. VAN FLANDERN in CRAIG – SMITH pp. 216-220.228.

the traditional time with something else that in a deceptive way is called time as the traditional time.

Therefore, if we want to continue to use the two Einsteinian Theories of Relativity instead of two analogous Lorentzian Theories, we should never use the name time in a deceptive way in the two Einsteinian Theories of Relativity so that we never confuse something else with the traditional time. In fact it is exactly this deceptive use that has deceived the academic physicists, so that they have erroneously asserted the truth of the relativistic temporal paradoxes.

In conclusion, both in the case that we use the two Einsteinian Theories of Relativity and in the case that we use instead two analogous Lorentzian Theories, all the temporal paradoxes in the field of relativistic physics are completely false.

GENERAL CONCLUSION: ALL THE RELATIVISTIC TEMPORAL PARADOXES ARE COMPLETELY FALSE

Therefore all the temporal paradoxes that are based on the two Einsteinian Theories of Relativity are based only on an incorrect interpretation of the same two Einsteinian Theories of Relativity. In fact these temporal paradoxes are based only on erroneous definitions of time, even in contrast with the same Einsteinian Theories of Relativity: as we have seen, once we use the relativisticly correct definitions of time the temporal paradoxes vanishes completely. In other words, as we have seen, only the incorrect definitions of time produce the relativistic temporal paradoxes, but in this case the relativistic temporal paradoxes are relative to a pseudo-time, that is not a true time, and therefore there are no true temporal paradoxes in the field of application of the two Einsteinian Theories of Relativity.

On the other hand, it had to be already completely clear that the time, as it was traditionally understood, cannot be influenced in any way by velocities of objects and/or gravitational fields, since this time is completely independent both from any velocity of physical bodies and from any physical force. Therefore the temporal paradoxes that are based only on velocities of objects (Special Theory of Relativity) and/or gravitational fields (General Theory of Relativity) are clearly based only on gross mistakes.

Moreover, as we have seen, the possibility of using two Lorentzian Theories instead of the two Einsteinian Theories of

Relativity entails that all the temporal paradoxes in the field of application of relativistic physics are completely false.

In conclusion, the entire community of the academic physicists has been making many blunders for over 100 years! Therefore, I hope that this publication can stop this scandalous fact in a short period of time.

BIBLIOGRAPHY

ANDERSON J. L., *Principles of Relativity Physics*, Academic Press, New York 1967.

CRAIG W. L., *Time and the Metaphysics of Relativity*, Kluwer Academic Publishers, Dordrecht 2001.

CRAIG W. L. – SMITH Q. (eds.), *Einstein, Relativity and Absolute Simultaneity*, Routledge, London 2007.

DAVIES P., *I misteri del tempo*, Arnoldo Mondadori Editore, Milano 1997 {Italian translation of DAVIES P., *About Time*, Orion Productions, Adelaide 1995}.

EISBERG R. – RESNICK R., *Quantum Physics of Atoms, Molecules, Solids, Nuclei, and Particles*, John Wiley & Sons, New York 1985^2.

EVERETT A. – ROMAN T., *Come viaggeremo nel tempo*, il Saggiatore, Milano 2016 {Italian translation of EVERETT A. – ROMAN T., *Time Travel*, The University of Chicago Press, Chicago 2012}.

GOTT J. R. III, *Viaggiare nel tempo*, Arnoldo Mondadori Editore, Milano 2002 {Italian translation of GOTT J. R. III, *Time Travel in Einstein's Universe*, Houghton Mifflin Harcourt, Boston 2002}.

HAWKING S., *Dal Big Bang ai Buchi Neri*, Biblioteca Universale Rizzoli, Milano 1997 {Italian translation of HAWKING S., *A Brief History of Time*, Bantam Dell Publishing Group, New York 1988}.

LORENTZ H. A. – EINSTEIN A. – MINKOWSKI H. – WEYL H., *The Principle of Relativity*, Dover Publications, New York 1952.

MISNER C. W. – THORNE K. S. – WHEELER J. A., *Gravitation*, W. H. Freeman and Company, New York 1973.

OHANIAN H. C. – RUFFINI R., *Gravitation and Spacetime*, Cambridge University Press, New York 2013^3.

PIERCE J. R., *Relativity and Space Travel*, Proc. IRE 47, (1959) 1053-1061.

RINDLER W., *Essential Relativity. Special, General, and Cosmological*, Springer-Verlag, New York 1979^2.

WEINBERG S., *Gravitation and Cosmology: Principles and Applications of the General Theory of Relativity*, John Wiley & Sons, New York 1972.

TABLE OF CONTENTS

Finito di stampare nel mese di Giugno 2016
per conto di Youcanprint *Self-Publishing*